FUN FACT FILE: ENGINEERING MARVELS

20 FUN FACTS ABOUT FAMOUS ROADS AND HIGHWAYS

BY KATE LIGHT

Gareth Stevens PUBLISHING

Please visit our website, www.garethstevens.com. For a free color catalog of all our high-quality books, call toll free 1-800-542-2595 or fax 1-877-542-2596.

Cataloging-in-Publication Data

Names: Light, Kate.
Title: 20 fun facts about roads and highways / Kate Light.
Description: New York : Gareth Stevens Publishing, 2020. | Series: Fun fact file: engineering marvels | Includes glossary and index.
Identifiers: ISBN 9781538246627 (pbk.) | ISBN 9781538246641 (library bound) | ISBN 9781538246634(6 pack)
Subjects: LCSH: Roads–Juvenile literature. | Roads–Design and construction–Juvenile literature. | Express highways–Juvenile literature.
Classification: LCC TE149.L49 2020 | DDC 388.1–dc23

First Edition

Published in 2020 by
Gareth Stevens Publishing
111 East 14th Street, Suite 349
New York, NY 10003

Designer: Sarah Liddell
Editor: Therese Shea

Photo credits: Cover, p. 1 (main) Steve Lagreca/Shutterstock.com; file folder used throughout David Smart/Shutterstock.com; binder clip used throughout luckyraccoon/Shutterstock.com; wood grain background used throughout ARENA Creative/Shutterstock.com; p. 5 trekandshoot/Shutterstock.com; p. 6 Laurens~commonswiki/Wikimedia Commons; p. 7 karamysh/Shutterstock.com; p. 8 Maiquel Jantsch/Shutterstock.com; p. 9 rest/Shutterstock.com; p. 10 Tm/Wikimedia Commons; p. 11 Lukas Bischoff Photograph/Shutterstock.com; p. 12 terekhov igor/Shutterstock.com; p. 13 Robert Biedermann/Shutterstock.com; p. 14 byvalet/Shutterstock.com; p. 15 travelview/Shutterstock.com; p. 16 Andrey Armyagov/Shutterstock.com; p. 17 Tupungato/Shutterstock.com; p. 18 Maks Ershov/Shutterstock.com; p. 19 Lynne Gilbert/Contributor/Moment Mobile/Getty Images; p. 20 Makicom/Shutterstock.com; p. 21 BedaNo1/Wikimedia Commons; p. 22 Aliciak3yz/Shutterstock.com; p. 23 Fletcher6/Wikimedia Commons; p. 24 A G Baxter/Shutterstock.com; p. 25 Mariusz Szczygiel/Shutterstock.com; p. 26 Andreas Jung/Shutterstock.com; p. 27 DigitalGlobe/Contributor/DigitalGlobe/Getty Images; p. 29 Claudio Divizia/Shutterstock.com.

Printed in the United States of America

CPSIA compliance information: Batch #CW20GS: For further information contact Gareth Stevens, New York, New York at 1-800-542-2595.

CONTENTS

Words in the glossary appear in **bold** type the first time they are used in the text.

MAKING CONNECTIONS

Roads are **engineering** wonders that have connected places and people throughout history. Ancient European and Asian peoples started to trade goods and share ideas because of the Silk Road. This was a **network** of routes, or paths, that connected Europe and Asia from 130 BC to AD 1453.

The famous Route 66 **highway** also connected east and west—in the United States. It ran along small towns from Chicago, Illinois, to Santa Monica, California. What other interesting roads have shaped history? Let's find out!

Lots of weird and wonderful **tourist attractions** sprang up along Route 66, making it popular for road trips.

OLDEST ROADS

FUN FACT: 1

HUMANS MAY HAVE USED A "ROAD" TO TRAVEL TO NORTH AMERICA FROM ASIA!

Many scientists think people walked from today's Siberia to Alaska more than 13,000 years ago! The route was called the Bering Land Bridge, or Beringia. It's now covered by seawater.

The idea that humans crossed the Bering Land Bridge is a theory, or an explanation based on facts that is generally accepted by scientists.

The Old North Trail is important to the Blackfeet people. It ran from Canada to Mexico, along the Rocky Mountains. It took 4 years to walk the whole way!

FUN FACT: 2

THERE'S A 10,000-YEAR-OLD ROAD IN NORTH AMERICA CALLED THE OLD NORTH TRAIL.

This path was created by native peoples—possibly including the humans who crossed the Bering Land Bridge. People first traveled by foot, then with dogs pulling goods, and then with horses.

INCA ENGINEERS

FUN FACT: 3

THE INCA NETWORK OF ROADS, STAIRWAYS, AND BRIDGES WAS ABOUT 25,000 MILES (40,234 KM) LONG.

The Inca, who trace back to the 1100s, created roads of dirt, grass, or sand. The routes were sometimes lined with stone walls or wooden posts. The Inca also built drains to prevent flooding.

You can still walk on roads and bridges built by the Inca near Machu Picchu, an ancient city in the Andes Mountains.

INNOVATIONS ON THE INCA ROADS

FEATURE:	DEFINITION:
SUSPENSION BRIDGES	BRIDGES CONSTRUCTED OF ROPES MADE FROM PLANTS
OROYA	HANGING BASKETS THAT COULD CARRY SEVERAL PEOPLE OVER DEEP VALLEYS
REST STATION	PLACES FOR TRAVELERS TO TAKE A BREAK ALONG THE ROAD
CHASKI OR *CHASQUIS*	RUNNERS WHO CARRIED GOODS AND MESSAGES ALONG THE ROADS
QUIPU	CORDS WITH SMALLER STRINGS AND **KNOTS** USED TO HELP RUNNERS REMEMBER MESSAGES

ALL ROADS LEAD TO ROME

FUN FACT: 4

ROMAN SOLDIERS ENGINEERED ANCIENT ROADS.

The Roman armies built about 50,000 miles (80,400 km) of roads. These connected the city of Rome to the lands the **empire** conquered, or took over. That might be where the saying "all roads lead to Rome" comes from.

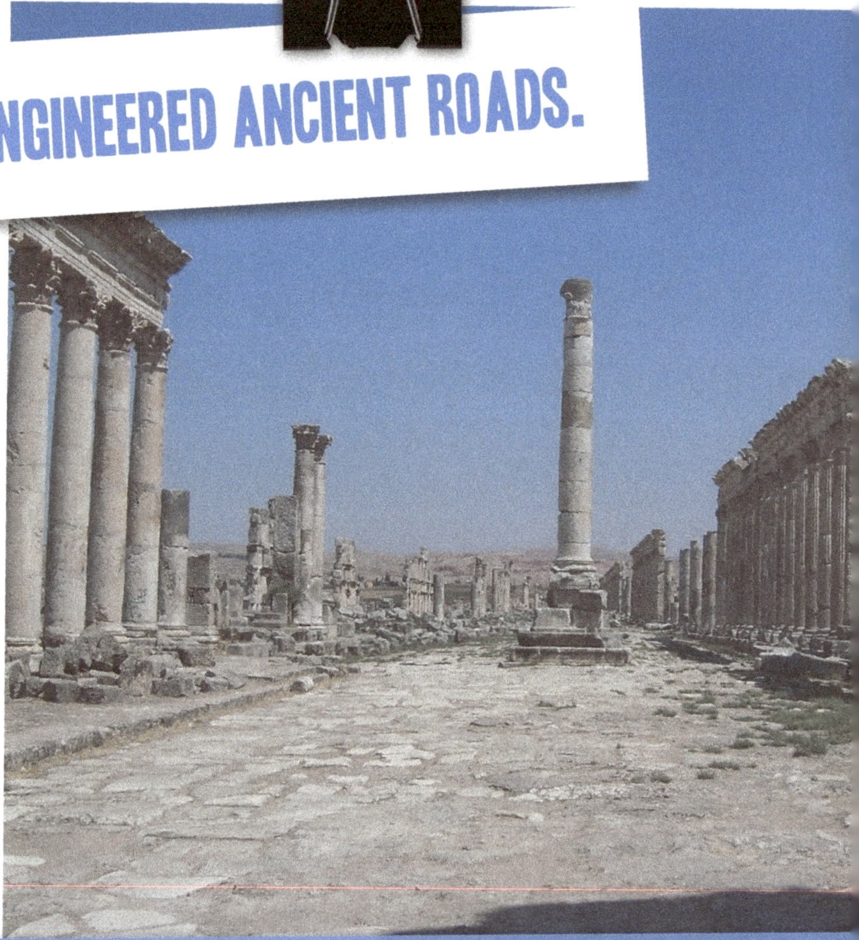

Many Roman cities had a *cardo maximus*, a **paved** road that ran from north to south. This picture shows the remains of a *cardo maximus* in Apamea, Syria.

The King's Highway, which ran through what is now Jordan, is still a tourist attraction.

FUN FACT: 5

THE KING'S HIGHWAY IS SO OLD IT'S TALKED ABOUT IN THE OLD TESTAMENT OF THE BIBLE!

This route ran from today's Syria to the north of the Red Sea. Trajan, who ruled the Roman empire from AD 98 to 117, renamed King's Highway *Via Nova Traiana*, or "Trajan's New Road."

FUN FACT: 6

THE LONGEST HIGHWAY IN THE WORLD NEARLY CONNECTS ALASKA TO ARGENTINA!

The Pan-American Highway isn't one road, but a network of roads. However, there's a 50-mile (80 km) gap in it in Panama. Travelers take a boat or plane around this forested area to continue on.

The Pan-American Highway is nearly 30,000 miles (48,300 km) long.

Kangaroos and other Australian animals can be spotted along Highway 1, shown on this map as a blue line.

FUN FACT: 7

YOU CAN DRIVE ALONG THE EDGE OF THE ENTIRE CONTINENT OF AUSTRALIA ON HIGHWAY 1!

Driving all 9,010 miles (14,500 km) of Highway 1 is called doing the "Big Lap." It's one of the longest road routes in the world and connects most major Australian cities.

FUN FACT: 8

BALDWIN STREET IN DUNEDIN, NEW ZEALAND, IS CALLED THE STEEPEST STREET IN THE WORLD.

Baldwin Street is **concrete** instead of tar, which is the sticky black matter made of coal often used for roads. Tar might melt and slide down the steep street in the hot summer weather!

The houses on Baldwin Street look odd because of the steep angle.

Building a highway through mountains is dangerous work. Nearly 900 workers died while building the Karakoram Highway.

FUN FACT: 9

THE WORLD'S HIGHEST PAVED HIGHWAY RISES 3 MILES (4.8 KM) ABOVE SEA LEVEL!

China and Pakistan built the Karakoram Highway to connect their countries. It goes through several mountain ranges, including the Himalayas, and is about 800 miles (1,287 km) long.

WEIRDEST WAYS

FUN FACT: 10

THE ATLANTIC ROAD IN NORWAY CAN FEEL A BIT LIKE A ROLLER COASTER!

This road, also called the Atlantic Ocean Road, connects an archipelago, or chain of islands, and includes eight bridges. It has many twists, turns, and dips. Sometimes, waves crash over parts of it!

There are stopping points along the Atlantic Road for people to take pictures of the view. You can also bike it!

Engineers **designed** the Hanshin Expressway to stand on large posts, so it's not actually touching the building. An expressway is a kind of high-speed highway.

FUN FACT: 11

IN JAPAN, YOU CAN DRIVE INTO A BUILDING—AND OUT THE OTHER SIDE!

Some roads go under buildings. However, the Hanshin Expressway travels through the fifth, sixth, and seventh floors of the Gate Tower Building in Osaka, Japan!

CALIFORNIA DREAMIN'

FUN FACT: 12

A SAN FRANCISCO STREET IS FAMOUS FOR EIGHT ZIGZAG TURNS—ALL IN ONE BLOCK.

Lombard Street looks scary, but the **switchbacks** were designed for safety. They help cars drive down the slope. It would be too dangerous to drive straight down!

Although other crooked, or twisted, streets claim the title, Lombard Street is called "the crookedest street in the world."

This musical road was originally engineered for a car **commercial**!

FUN FACT: 13

YOU CAN MAKE MUSIC BY DRIVING ALONG THE CIVIC MUSICAL ROAD IN LANCASTER, CALIFORNIA!

The left lane has many grooves, or narrow cuts, in it. When a car's tires go over them at a certain speed, the noises sound like part of the song called the "William Tell Overture."

RACING ROADS

FUN FACT: 14

THE TV SHOW *TOP GEAR* NAMED ITALY'S STELVIO PASS THE "GREATEST DRIVING ROAD IN THE WORLD."

This road through the Italian Alps is known for dozens of tight turns. People ride bicycles, motorcycles, and cars on the pass for an exciting ride!

Would you dare to take a ride up and down these killer switchbacks?

Racetracks are designed to fit the features of the land. They need to be exciting, but also safe. It takes careful engineering to make a great track.

FUN FACT: 15

WHEN YOU LEARN TO DRIVE, YOU CAN DRIVE ON A REAL RACETRACK IN GERMANY!

The Nürburgring racetrack is famous for 154 "corners," or turns, that scare even the pros. However, ordinary people can also take a spin at certain times of the day.

BIGGER IN TEXAS

FUN FACT: 16

THE WIDEST HIGHWAY IN THE WORLD IS 26 LANES ACROSS AT ONE POINT!

The Katy Freeway through Houston, Texas, had too much **traffic**, so it was made much wider. However, even 26 lanes couldn't solve the problem. Traffic continued to get worse!

In 2018, the average American spent 97 hours stuck in traffic!

This interchange is 12 stories tall in some places. An interchange is a place where highways meet using a system with different levels.

FUN FACT: 17

THE HIGH FIVE INTERCHANGE IS—YOU GUESSED IT!—FIVE LEVELS HIGH.

When you can't build wider roads, build them higher! Engineers designed the High Five Interchange in Dallas, Texas, for the 500,000 people who travel it each day.

THE MAGIC OF CIRCLES

FUN FACT: 18

ROUNDABOUTS ARE SAFER THAN STOP SIGNS OR STOP LIGHTS.

Roundabouts are a type of intersection, or a place where different roads meet. They're shaped like a circle. They make drivers slow down—and lower the number of crashes in which someone gets hurt by 75 percent!

The Magic Roundabout in Swindon, England, is an intersection made up of six roundabouts!

AROUND THE ROUNDABOUT

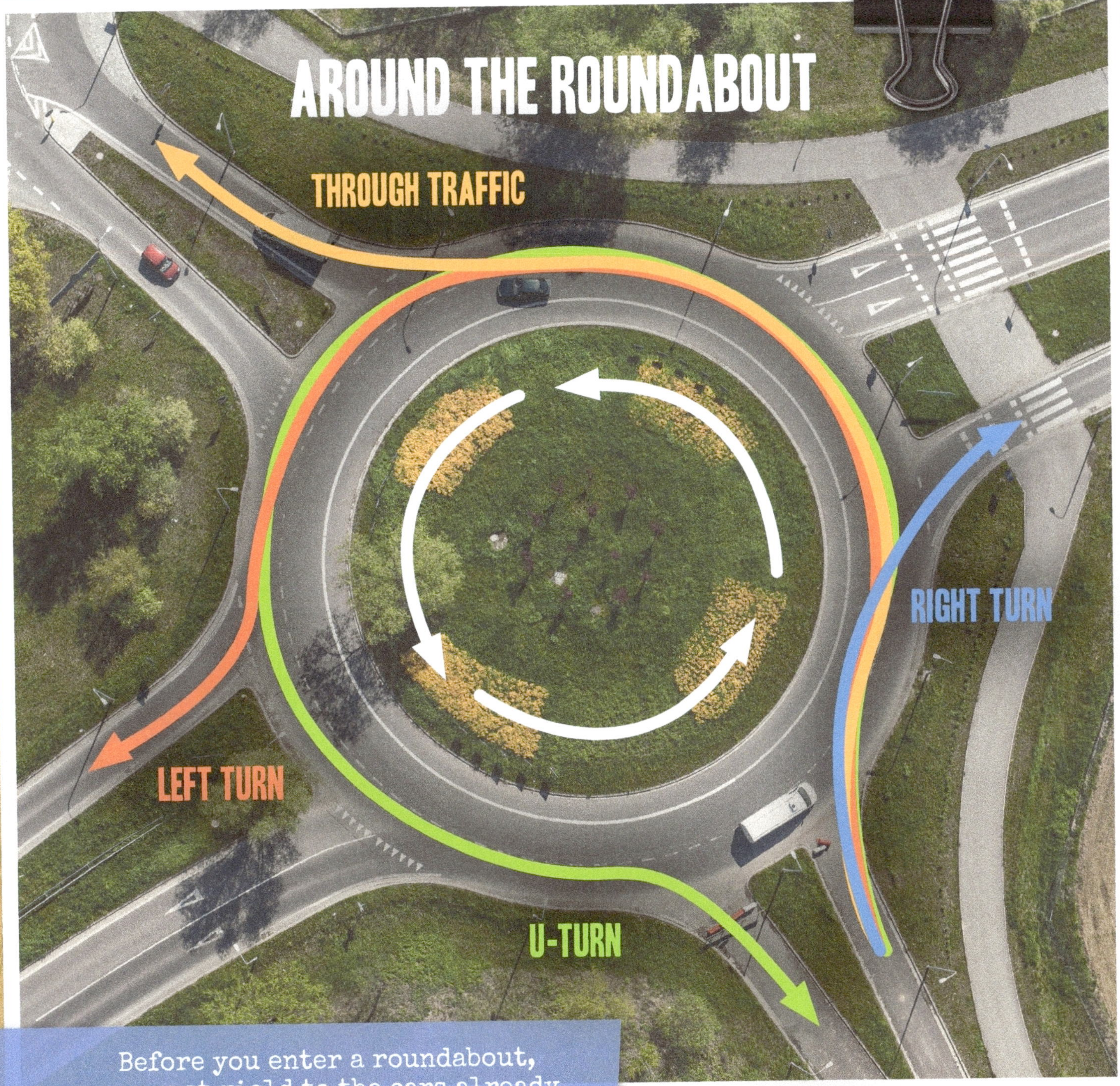

Before you enter a roundabout, you must yield to the cars already driving around the circle. This means you have to wait your turn!

ROAD VS. NATURE

FUN FACT: 19

THE PASSAGE DU GOIS IN FRANCE IS A ROAD THAT ONLY APPEARS TWICE A DAY—AT LOW TIDE!

The Passage du Gois connects the island of Noirmoutier to mainland France, but just for a couple of hours each day. Most of the time, it's under several feet of water!

There are special poles along the Passage du Gois for anyone who's only part way across when the tide comes in. They have to sit and wait for low tide—or for a boat!

The Tibbitt to Contwoyto Winter Road is as wide as an eight-lane highway at some points. You can see mines and some of the route here.

FUN FACT: 20

THE TIBBITT TO CONTWOYTO WINTER ROAD IS BUILT OVER FROZEN LAKES EVERY YEAR IN NORTHERN CANADA.

Mining companies have been building this ice road since 1982. Trucks drive across the road to mines, then carry heavy loads back. The road is about 370 miles (595 km) long.

LIFE IS A HIGHWAY

Roads are still shaping life in big and small ways today. Some are even destinations, or places to visit, themselves.

One of the most famous roads in the world is Abbey Road in London, England. On the cover of The Beatles's 1969 album *Abbey Road*, the rock band is pictured on this road's crosswalk. Visitors come to take pictures of themselves walking where The Beatles once walked. They often have to wait for their turn. After all, the road is also for cars!

Plenty of roads are famous for one reason or another. Which road would you like to visit?

GLOSSARY

commercial: a break during a TV or radio show that tries to sell something

concrete: a hard, strong matter used for building which is made by mixing cement, sand, and rocks with water

continent: one of Earth's seven great landmasses

design: to create the pattern or shape of something

empire: a large area of land under the control of a single ruler

engineering: the use of science and math to build better objects

highway: a main road, especially one between towns and cities

knot: a tied piece of a rope or string

network: a system of things, such as roads, that are connected to each other

paved: covered with a firm matter to make travel easier

switchback: a very sharp turn used in roads to help climb or go down a steep hill

tourist attraction: a place where people travel to visit

traffic: all the cars, trucks, and other vehicles driving along a certain road or in a certain area

FOR MORE INFORMATION

BOOKS

Frankopan, Peter. *The Silk Roads*. New York, NY: Bloomsbury Children's Books, 2018.

Krajnik, Elizabeth. *Paving Roads and Highways*. New York, NY: PowerKids Press, 2019.

Snedden, Robert. *Roads*. New York, NY: PowerKids Press, 2017.

WEBSITES

How Do You Build a Road?
www.wonderopolis.org/wonder/how-do-you-build-a-road
Find out how engineers build the roads you need.

The Silk Road
www.ducksters.com/history/china/silk_road.php
Learn about the ancient Silk Road at this cool site.

Publisher's note to educators and parents: Our editors have carefully reviewed these websites to ensure that they are suitable for students. Many websites change frequently, however, and we cannot guarantee that a site's future contents will continue to meet our high standards of quality and educational value. Be advised that students should be closely supervised whenever they access the internet.

INDEX